John Dalton, And others

Foundations of the Molecular Theory

John Dalton, And others

Foundations of the Molecular Theory

ISBN/EAN: 9783337059613

Printed in Europe, USA, Canada, Australia, Japan

Cover: Foto ©berggeist007 / pixelio.de

More available books at **www.hansebooks.com**

Alembic Club Reprints—No. 4.

FOUNDATIONS

OF THE

MOLECULAR THEORY:

COMPRISING

PAPERS AND EXTRACTS

BY

JOHN DALTON,

JOSEPH-LOUIS GAY-LUSSAC,

AND

AMEDEO AVOGADRO,

(1808–1811.)

EDINBURGH:
WILLIAM F. CLAY, 18 TEVIOT PLACE.
LONDON:
SIMPKIN, MARSHALL, HAMILTON, KENT, & CO. LTD.
1893.

PREFACE.

THE papers here reprinted in chronological order serve to exhibit the historical development of the idea of a connection existing between the number of particles in different gases and the volume they occupy.

It will be seen that Dalton from the first entertains the notion that equal volumes of different gases may contain the same number of ultimate particles at equal temperature and pressure, but that he is legitimately forced to reject this assumption, conceiving no distinction between the atom and the molecule of an element. Gay-Lussac's important experimental work on the combining volumes of gases then shows the necessity of a simple relation between the ultimate particles of gases and their volumes, although he does not point this out in his paper. Dalton, however, perceives the necessity, and characteristically concludes by doubting the accuracy of Gay-Lussac's experiments. Avogadro, finally, accepts both Dalton's theory and Gay-Lussac's data, and teaches how to reconcile them by distinguishing between the atom and the molecule of an elementary gas.

It has not been thought necessary to reprint the letter of Ampère to Berthollet (Annales de Chimie, 90, 43-86, 1814), since that paper contains no advance on the views of Avogadro published three years earlier, its author simply drawing the same conclusions from the same premises.

The English version of the French originals will probably be found more faithful than elegant, especially so in the case of Avogadro's paper, where the French is always clumsy and occasionally obscure.

J. W.

FOUNDATIONS OF THE MOLECULAR THEORY.

1. EXTRACTS FROM A NEW SYSTEM OF CHEMICAL PHILOSOPHY. PART I. (1808). BY JOHN DALTON.

I.

(pp. 70–72.)

3. *The quantity of heat belonging to the ultimate particles of all elastic fluids, must be the same under the same pressure and temperature.*

It is evident the number of ultimate particles or molecules in a given weight or volume of one gas is not the same as in another : for, if equal measures of azotic and oxygenous gases were mixed, and could be instantly united chemically, they would form nearly two measures of nitrous gas, having the same weight as the two original measures ; but the number of ultimate particles could at most be one half of that before the union. No two elastic fluids, probably, therefore, have the same number of particles, either in the same volume or the same weight. Suppose, then, a given volume of any elastic fluid to be constituted of particles, each surrounded with an atmosphere of heat, repelling each other through the medium of those atmospheres, and in a state of equilibrium under the pressure of a constant force, such as the earth's atmosphere, also at the temperature of the surrounding bodies; suppose further, that by some sudden change each molecule of air was endued with a stronger affinity for heat; query the change that would take place in consequence of this last supposition? The only answer that can be given, as it appears to me, is this.— The particles will condense their respective atmospheres of heat, by which their mutual repulsion will be diminished, and the external pressure will therefore effect a proportionate condensation in the volume of air: neither an increase nor diminution in the quantity of heat around each molecule, or around the whole, will take place. Hence the truth of the supposition, or as it may now be called, proposition, is demonstrated.

II.

(pp. 187-189.)

I shall now proceed to give my present views on the subject of mixed gases, which are somewhat different from what they were when the theory was announced, in consequence of the fresh lights which succeeding experience has diffused. In prosecuting my enquiries into the nature of elastic fluids, I soon perceived it was necessary, if possible, to ascertain whether the atoms or ultimate particles of the different gases are of the same size or volume in like circumstances of temperature and pressure. By the size or volume of an ultimate particle, I mean in this place, the space it occupies in the state of a pure elastic fluid; in this sense the bulk of the particle signifies the bulk of the supposed impenetrable nucleus, together with that of its surrounding repulsive atmosphere of heat. At the time I formed the theory of mixed gases, I had a confused idea, as many have, I suppose, at this time, that the particles of elastic fluids are all of the same size ; that a given volume of oxygenous gas contains just as many particles as the same volume of hydrogenous ; or if not, that we had no data from which the question could be solved. But from a train of reasoning, similar to that exhibited at page 71, I became convinced that different gases have *not* their particles of the same size : and that the following may be adopted as a maxim, till some reason appears to the contrary : namely,—

That every species of pure elastic fluid has its particles, globular and all of a size ; but that no two species agree in the size of their particles, the pressure and temperature being the same.

2. MEMOIR ON THE COMBINATION OF GASEOUS SUBSTANCES WITH EACH OTHER. BY M. GAY-LUSSAC.*

Read before the Philomathic Society, 31st December 1808.

———————><———————

SUBSTANCES, whether in the solid, liquid, or gaseous state, possess properties which are independent of the force of cohesion ; but they also possess others which appear to be modified by this force (so variable in its intensity), and which no longer follow any regular law. The same pressure applied to all solid or liquid substances would produce a diminution of volume differing in each case, while it would be equal for all elastic fluids. Similarly, heat expands all substances; but the dilatations of liquids and solids have hitherto presented no regularity, and it is only those of elastic fluids which are equal and independent of the nature of each gas. The attraction of the molecules in solids and liquids is, therefore, the cause which modifies their special properties ; and it appears that it is only when the attraction is entirely destroyed, as in gases, that bodies under similar conditions obey simple and regular laws. At least, it is my intention to make known some new properties in gases, the effects of which are regular, by showing that these substances combine amongst themselves in very simple proportions, and that the contraction of volume which they experience on combination also follows a

---

* Mémoires de la Société d'Arcueil, II. (1809), pp. 207-234.

regular law. I hope by this means to give a proof of an idea advanced by several very distinguished chemists— that we are perhaps not far removed from the time when we shall be able to submit the bulk of chemical pheno- mena to calculation.

It is a very important question in itself, and one much discussed amongst chemists, to ascertain if compounds are formed in all sorts of proportions. M. Proust, who appears first to have fixed his attention on this subject, is of opinion that the metals are susceptible of only two de- grees of oxidation, a *minimum* and a *maximum ;* but led away by this seductive theory, he has seen himself forced to entertain principles contrary to physics in order to reduce to two oxides all those which the same metal sometimes presents. M. Berthollet thinks, on the other hand— reasoning from general considerations and his own ex- periments—that compounds are always formed in very variable proportions, unless they are determined by special causes, such as crystallisation, insolubility, or elasticity. Lastly, Dalton has advanced the idea that compounds of two bodies are formed in such a way that one atom of the one unites with one, two, three, or more atoms of the other.* It would follow from this mode of looking at compounds that they are formed in constant proportions, the existence of intermediate bodies being excluded, and in this respect Dalton's theory would resemble that of M. Proust; but M. Berthollet has already strongly opposed it in the Introduction he has written to Thomson's Chemistry, and we shall see that in reality it is not entirely exact. Such is the state of the

---

* Dalton has been led to this idea by systematic considerations ; and one may see from his work, A New System of Chemical Philosophy, p. 213, and from that of Thomson, Vol. 6, that his researches have no connection with mine.

question now under discussion ; it is still very far from receiving its solution, but I hope that the facts which I now proceed to set forth, facts which had entirely escaped the notice of chemists, will contribute to its elucidation.

Suspecting, from the exact ratio of 100 of oxygen to 200 of hydrogen, which M. Humboldt and I had determined for the proportions of water, that other gases might also combine in simple ratios, I have made the following experiments. I prepared fluoboric,* muriatic, and carbonic gases, and made them combine successively with ammonia gas. 100 parts of muriatic gas saturate precisely 100 parts of ammonia gas, and the salt which is formed from them is perfectly neutral, whether one or other of the gases is in excess. Fluoboric gas, on the contrary, unites in two proportions with ammonia gas. When the acid † gas is put first into the graduated tube, and the other gas is then passed in, it is found that equal volumes of the two condense, and that the salt formed is neutral. But if we begin by first putting the ammonia gas into the tube, and then admitting the fluoboric gas in single bubbles, the first gas will then be in excess with regard to the second, and there will result a salt with excess of base, composed of 100 of fluoboric gas and 200 of ammonia gas. If carbonic gas is brought into contact with ammonia gas, by passing it sometimes first, sometimes second into the tube, there is always formed a sub-carbonate composed of 100 parts of carbonic gas and 200 of ammonia gas. It may, however, be proved that neutral carbonate of ammonia would be composed of equal volumes of each of these components. M. Ber-

---

* M. Thenard and I have given the name of fluoboric gas to that particular gas which we obtained by distilling pure fluoride of lime with vitreous boracic acid.

† ["Alkaline" in the original.]

thollet, who has analysed this salt, obtained by passing
carbonic gas into the sub-carbonate, found that it was
composed of 73.34 parts by weight of carbonic gas and
26.66 of ammonia gas. Now, if we suppose it to be
composed of equal volumes of its components, we find
from their known specific gravity, that it contains by
weight

71.81 of carbonic acid,
28.19 of ammonia,
———
100.0

a proportion differing only slightly from the preceding.

If the neutral carbonate of ammonia could be formed
by the mixture of carbonic gas and ammonia gas, as
much of one gas as of the other would be absorbed; and
since we can only obtain it through the intervention of
water, we must conclude that it is the affinity of this
liquid which competes with that of the ammonia to over-
come the elasticity of the carbonic acid, and that the
neutral carbonate of ammonia can only exist through the
medium of water.

Thus we may conclude that muriatic, fluoboric, and
carbonic acids take exactly their own volume of ammonia
gas to form neutral salts, and that the last two take twice
as much to form *sub-salts*. It is very remarkable to see
acids so different from one another neutralise a volume
of ammonia gas equal to their own; and from this we
may suspect that if all acids and all alkalis could be
obtained in the gaseous state, neutrality would result
from the combination of equal volumes of acid and
alkali.

It is not less remarkable that, whether we obtain a
neutral salt or a *sub-salt*, their elements combine in simple
ratios which may be considered as limits to their proportions. Accordingly, if we accept the specific gravity of

muriatic acid determined by M. Biot and myself,* and
those of carbonic gas and ammonia given by MM. Biot
and Arago, we find that dry muriate of ammonia is
composed of

$$\begin{array}{lll} \text{Ammonia,} & \text{100.0} & \text{38.35} \\ \text{Muriatic acid,} & \text{160.7} ^{\text{or}} & \text{61.65} \\ & & \overline{} \\ & & \text{100.00} \end{array}$$

a proportion very far from that of M. Berthollet—

100 of ammonia,
213 of acid.

In the same way, we find that sub-carbonate of
ammonia contains

$$\begin{array}{lll} \text{Ammonia,} & \text{100.0} & \text{43.98} \\ \text{Carbonic acid,} & \text{127.3} ^{\text{or}} & \text{56.02} \\ & & \overline{} \\ & & \text{100.00} \end{array}$$

and the neutral carbonate

$$\begin{array}{lll} \text{Ammonia,} & \text{100.0} & \text{28.19} \\ \text{Carbonic acid,} & \text{254.6} ^{\text{or}} & \text{71.81} \\ & & \overline{} \\ & & \text{100.00} \end{array}$$

It is easy from the preceding results to ascertain the
ratios of the capacity of fluoboric, muriatic, and carbonic
acids; for since these three gases saturate the same
volume of ammonia gas, their relative capacities will be
inversely as their densities, allowance having been made
for the water contained in muriatic acid.

We might even now conclude that gases combine with
each other in very simple ratios; but I shall still give
some fresh proofs.

---

* As muriatic gas contains one-fourth its weight of water, we
must only take three-fourths of the density for that of real muriatic
acid.

According to the experiments of M. Amédée Berthollet, ammonia is composed of

> 100 of nitrogen,
> 300 of hydrogen,

by volume.

I have found (1st vol. of the Société d'Arcueil) that sulphuric acid is composed of

> 100 of sulphurous gas,
> 50 of oxygen gas.

When a mixture of 50 parts of oxygen and 100 of carbonic oxide· (formed by the distillation of oxide of zinc with strongly calcined charcoal) is inflamed, these two gases are destroyed and their place taken by 100 parts of carbonic acid gas. Consequently carbonic acid may be considered as being composed of

> 100 of carbonic oxide gas,
> 50 of oxygen gas.

Davy, from the analysis of various compounds of nitrogen with oxygen, has found the following proportions by weight :—

	Nitrogen.	Oxygen.
Nitrous oxide - - -	63.30	36.70
Nitrous gas - - - -	44.05	55.95
Nitric acid - - - -	29.50	70.50

Reducing these proportions to volumes we find—

	Nitrogen.	Oxygen.
Nitrous oxide - - -	100	49.5
Nitrous gas - - -	100	108.9
Nitric acid - - -	100	204.7

The first and the last of these proportions differ only slightly from 100 to 50, and 100 to 200 ; it is only the second which diverges somewhat from 100 to 100. The difference, however, is not very great, and is such as we might expect in experiments of this sort ; and I have assured myself that it is actually nil. On burning the

new combustible substance from potash in 100 parts by
volume of nitrous gas, there remained over exactly 50
parts of nitrogen, the weight of which, deducted from
that of the nitrous gas (determined with great care by M.
Bérard at Arcueil), yields as result that this gas is composed of equal parts by volume of nitrogen and oxygen.

We may then admit the following numbers for the proportions by volume of the compounds of nitrogen with
oxygen :—

				Nitrogen.	Oxygen.
Nitrous oxide	-	-	-	100	50
Nitrous gas	-	-	-	100	100
Nitric acid	-	-	-	100	200

From my experiments, which differ very little from
those of M. Chenevix, oxygenated muriatic acid * is composed by weight of

Oxygen -	-	-	22.92
Muriatic acid		-	77.08

Converting these quantities into volumes, we find that
oxygenated muriatic acid is formed of

Muriatic gas -	-	300.0
Oxygen gas	-	-	103.2

a proportion very nearly

Muriatic gas -	-	300
Oxygen gas	-	-	100†

---

* [Chlorine.]

† In the proportion by weight of oxygenated muriatic acid, the
muriatic acid is supposed to be free from water, whilst in the proportion by volume it is supposed to be combined with a fourth of
its weight of water, which, since the reading of this paper, M.
Thenard and I have proved to be absolutely necessary for its existence in the gaseous state.   But since the simple ratio of 300 of acid
to 100 of oxygen cannot be due to chance, we must conclude that
water by combining with dry muriatic acid to form ordinary muriatic
acid does not sensibly change its specific gravity.   We should be
led to the same conclusion from the consideration that the specific

Thus it appears evident to me that gases always combine in the simplest proportions when they act on one another; and we have seen in reality in all the preceding examples that the ratio of combination is 1 to 1, 1 to 2, or 1 to 3. It is very important to observe that in considering weights there is no simple and finite relation between the elements of any one compound; it is only when there is a second compound between the same elements that the new proportion of the element that has been added is a multiple of the first quantity. Gases, on the contrary, in whatever proportions they may combine, always give rise to compounds whose elements by volume are multiples of each other.

Not only, however, do gases combine in very simple proportions, as we have just seen, but the apparent contraction of volume which they experience on combination has also a simple relation to the volume of the gases, or at least to that of one of them.

I have said, following M. Berthollet, that 100 parts of carbonic oxide gas, prepared by distilling oxide of zinc and strongly calcined charcoal, produce 100 parts of carbonic gas on combining with 50 of oxygen. It follows from this that the apparent contraction of the two gases is precisely equal to the volume of oxygen gas added. The density of carbonic gas is thus equal to that of carbonic oxide gas plus half the density of oxygen gas; or, conversely, the density of carbonic oxide gas is equal to that of carbonic gas, minus half that of oxygen gas. Accordingly, taking the density of air as unity, we find

gravity of oxygenated muriatic acid, which from our experiments contains no water, is exactly the same as that obtained by adding the density of oxygen gas to three times that of muriatic gas, and taking half of this sum. M. Thenard and I have also found that oxygenated muriatic gas contains precisely half its volume of oxygen, and that it can destroy in consequence its own volume of hydrogen.

the density of carbonic oxide gas to be 0.9678, instead of
0.9569 experimentally determined by Cruickshanks.
We know, besides, that a given volume of oxygen produces an equal volume of carbonic acid; consequently
oxygen gas doubles its volume on forming carbonic oxide
gas with carbon, and so does carbonic gas on being
passed over red-hot charcoal.　Since oxygen produces an
equal volume of carbonic gas, and the density of the
latter is well known, it is easy to calculate the proportion
of its elements.　In this way we find that carbonic gas
is composed of

> 27.38 of carbon,
> 72.62 of oxygen,

and carbonic oxide of

> 42.99 of carbon,
> 57.01 of oxygen.

Pursuing a similar course, we find that if sulphur takes
100 parts of oxygen to produce sulphurous acid, it takes
150 parts to produce sulphuric acid.　As a matter of
fact, we find that sulphuric acid, according to the experiments of MM. Klaproth, Bucholz, and Richter, is composed of 100 parts by weight of sulphur and 138 of
oxygen.

On the other hand sulphuric acid is composed of 2
parts by volume of sulphurous gas, and 1 of oxygen gas.
Consequently the weight of a certain quantity of sulphuric
acid should be the same as that of 2 parts of sulphurous
acid and 1 of oxygen gas, *i.e.*, 2 × 2.265, plus 1.10359 =
5.63359; seeing that, according to Kirwan, sulphurous
gas weighs 2.265, the density of air being taken as unity.
But from the proportion of 100 of sulphur to 138 of
oxygen, this quantity contains 3.26653 of oxygen, and if
we subtract from it 1.10359 there will remain 2.16294 for
the weight of oxygen in 2 parts of sulphurous acid, or
1.08147 for the weight of oxygen contained in 1 part.

Now as this last quantity only differs by 2 per cent. from 1.10359, which represents the weight of 1 part of oxygen gas, it must be concluded that oxygen gas, in combining with sulphur to form sulphurous gas, only experiences a diminution of a fiftieth of its volume, and this would probably be nil if the data I have employed were more exact. On this last supposition, using Kirwan's value for the specific gravity of sulphurous gas, we should find that this acid is composed of

100.00 of sulphur,
95.02 of oxygen.

But if, adopting the preceding proportions for sulphuric acid, we allow, as appears probable, that 100 of sulphurous gas contain 100 of oxygen gas, and that 50 have still to be added to convert it into sulphuric acid, we shall obtain for the proportions in sulphurous acid

100.00 of sulphur,
92.0 of oxygen.

Its specific gravity calculated on the same suppositions, and referred to that of air, would be 2.30314, instead of 2.2650 as Kirwan found directly.*

---

* In order to remove these differences it would be necessary to make new experiments on the density of sulphurous gas, on the direct union of oxygen gas with sulphur to see if there is contraction, and on the union of sulphurous gas with ammonia gas. I have found, it is true, on heating cinnabar in oxygen gas, that 100 parts of this gas only produce 93 of sulphurous gas. It also appeared as if less sulphurous gas than ammonia gas was necessary to form a neutral salt. But as these experiments were not made under suitable conditions,—especially the last, which could only be made in presence of water, the sulphurous gas decomposing and precipitating sulphur immediately on being mixed with the ammonia gas,—I intend to repeat them and determine exactly all the conditions before drawing any conclusion from them. This is all the more necessary, as sulphurous gas can be used to analyse sulphuretted hydrogen gas, if its proportions are well known.

B

Phosphorus is very closely connected with sulphur, seeing that both have nearly the same specific gravity. Consequently phosphorus should take up twice as much oxygen to become phosphorous acid, as to pass from this state into phosphoric acid. Since the latter is composed, according to Rose, of

100.0 of phosphorus,

114.0 of oxygen,

it follows that phosphorous acid should contain

100.0 of phosphorus,

76.0 of oxygen.

We have seen that 100 parts of nitrogen gas take 50 parts of oxygen gas to form nitrous oxide, and 100 of oxygen gas to form nitrous gas. In the first case, the contraction is a little greater than the volume of oxygen added ; for the specific gravity of nitrous oxide, calculated on this hypothesis, is 1.52092, while that given by Davy is 1.61414. But it is easy to show, from some of Davy's experiments, that the apparent contraction is precisely equal to the volume of oxygen gas added. On passing the electric spark through a mixture of 100 parts of hydrogen and 97.5 of nitrous oxide the hydrogen is destroyed, and 102 parts of nitrogen remain, including that quantity which is almost always mixed with the hydrogen, and a little of the latter gas which has escaped combustion. The residue, after making all corrections, would be very nearly equal in volume to the nitrous oxide employed. Similarly, on passing the electric spark through a mixture of 100 parts of phosphuretted hydrogen and 250 of nitrous oxide, water and phosphoric acid are formed, and exactly 250 parts of nitrogen remain,— another evident proof that the apparent contraction of the elements of nitrous oxide is equal to the whole volume of oxygen added. From this circumstance, its specific gravity referred to that of air should be 1.52092.

The apparent contraction of the elements of nitrous gas appears, on the other hand, to be nil. If we admit, as I have shown, that it is composed of equal parts of oxygen and nitrogen, we find that its density, calculated on the assumption that there is no contraction, is 1.036, while that determined directly is 1.038.

Saussure found that the density of water vapour is to that of air as 10 is to 14. Assuming that the contraction of volume of the two gases is only equal to the whole volume of oxygen added, we find instead of this a ratio of 10 to 16. This difference, and the authority of a physicist so distinguished as Saussure, would seem to be enough to make us reject the assumption I have just made; but I shall mention several circumstances that render it very probable. Firstly, it has a very strong analogy in its favour; secondly, M. Tralès found by direct experiment that the ratio of the density of water-vapour to air is 10 to 14.5, instead of 10 to 14; thirdly, although we do not know very exactly the volume occupied by water on passing into the elastic state, we do know, from the experiments of Watt, that a cubic inch of water produces nearly a cubic foot of steam, *i.e.*, a volume 1728 times as great. Now, adopting Saussure's ratio, we find only 1488 for the volume occupied by water when it is converted into steam; but adopting the ratio of 10 to 16, we should have 1700.6. Finally, the refraction of water-vapour, calculated on the assumption of the ratio 10 to 14, is a little greater than the observed refraction; but that calculated from the ratio 10 to 16 is much more in harmony with the results of experiment. These, then, are the considerations which go to make the ratio 10 to 16 very probable.

Ammonia gas is composed of three parts by volume of

hydrogen and one of nitrogen, and its density compared to air is 0.596. But if we suppose the apparent contraction to be half of the whole volume, we find 0.594 for the density. Thus it is proved, by this almost perfect concordance, that the apparent contraction of its elements is precisely half the total volume, or rather double the volume of the nitrogen.

I have already proved that oxygenated muriatic gas is composed of 300 parts of muriatic gas and 100 of oxygen gas. Admitting that the apparent contraction of the two gases is half the whole volume, we find 2.468 for its density, and by experiment 2.470. I have also assured myself by several experiments that the proportions of its elements are such that it forms neutral salts with the metals. For example, if we pass oxygenated muriatic gas over copper, there is formed a slightly acid green muriate, and a little oxide of copper is precipitated, because the salt cannot be obtained perfectly neutral. It follows from this that in all the muriates, as in oxygenated muriatic acid, the acid reduced to volume is thrice the oxygen. It would be the same for carbonates and fluorides, the acids of which have for equal volumes the same saturation capacity as muriatic acid.

We see, then, from these various examples, that the contraction experienced by two gases on combination is in almost exact relation with their volume, or rather with the volume of one of them. Only very slight differences exist between the densities of compounds obtained by calculation and those given by experiment, and it is probable that, on undertaking new researches, we shall see them vanish entirely.

Recalling the great law of chemical affinity, that every combination involves an approximation of the elementary molecules, it is difficult to conceive why carbonic oxide gas should be lighter than oxygen. Indeed, that is the principal reason which has led M. Berthollet to assume the existence of hydrogen in this gas, and thus explain its low density. But it seems to me that the difficulty arises from supposing that the approximation of the elementary molecules is represented in gases by the diminution of volume which they suffer on combination. This supposition is not always true, and we might cite several gaseous combinations, the constituent molecules of which would be brought very close together, although there is not only no diminution of volume, but even a dilatation. Such, for example, is nitrous gas, whether we consider it as being formed directly from nitrogen and oxygen, or from nitrous oxide and oxygen. In the first case, there is no diminution of volume; and in the second, there would be dilatation, for 100 parts of nitrous oxide and 50 of oxygen would produce 200 of nitrous gas. We know too that carbonic gas represents an exactly equal volume of oxygen, and that the affinity which unites its elements is very powerful. Nevertheless, if we admitted an immediate relation between the condensation of the elements and the condensation of volume, we should conclude, contrary to experiment, that there is no condensation. Otherwise it would be necessary to suppose that if carbon were in the gaseous state it would combine in equal volumes (or in any other proportion) with oxygen, and that the apparent condensation would then be equal to the whole volume of the gaseous carbon. But if we make this supposition for carbonic acid, we may also make it for carbonic oxide, by assuming, for instance, that 100 parts of gaseous carbon would produce 100 parts of the gas on combining with 50 parts of oxygen. However it may stand with these suppositions, which only serve to make it conceivable that oxygen can produce a compound lighter than itself by combining with a solid substance, we must admit, as a truth founded on a great number of observations, that the condensation of the molecules of two combining substances, in particular of two gases, has no immediate relation to the condensation of volume, since we often see that whilst one is very great the other is very small or even nil.

The observation that the gaseous combustibles combine with oxygen in the simple ratios of 1 to 1, 1 to 2, 1 to $\frac{1}{2}$, can lead us to determine the density of the vapours of combustible substances, or at least to approximate closely to that determination. For if we suppose all combustible substances to be in the gaseous state, a specified volume of each would absorb an equal volume of oxygen, or twice as much, or else half; and as we know the proportion of oxygen taken up by each combustible substance in the solid or liquid state, it is sufficient to convert the oxygen into volumes and also the combustible, under the condition that its vapour shall be equal to the volume of oxygen, or else double or half this value. For example, mercury is susceptible of two degrees of oxidation, and we may compare the first one to nitrous oxide. Now, according to MM. Fourcroy and Thenard, 100 parts of mercury absorb 4.16, which reduced to gas would occupy a space of 8.20. These 100 parts of mercury reduced to vapour should therefore occupy twice the space, viz., 16.40. We thence conclude that the density of mercury vapour is 12.01 greater than that of oxygen, and that the metal on passing from the liquid to the gaseous state assumes a volume 961 times as great.

I shall not discuss more of these determinations, because they are only based on analogies, and it is besides easy to multiply them. I shall conclude this Memoir by examining if compounds are formed in constant or variable proportions, as the experiments of which I have just given an account lead me to the discussion of these two opinions.

According to Dalton's ingenious idea, that combinations are formed from atom to atom, the various compounds which two substances can form would be produced by the union of one molecule of the one with one molecule of the other, or with two, or with a greater number, but always without intermediate compounds. Thomson and Wollaston have indeed described experiments which appear to confirm this theory. Thomson * has found that super-oxalate of potash contains twice as much acid as is necessary to saturate the alkali; and Wollaston,† that the sub-carbonate of potash contains, on the other hand, twice as much alkali as is necessary to saturate the acid.

The numerous results I have brought forward in this Memoir are also very favourable to the theory. But M. Berthollet, who thinks that combinations are made continuously, cites in proof of his opinion the acid sulphates, glass, alloys, mixtures of various liquids,—all of which are compounds with very variable proportions, and he insists principally on the identity of the force which produces chemical compounds and solutions.

Each of these two opinions has, therefore, a large number of facts in its favour; but although they are apparently utterly opposed, it is easy to reconcile them.

We must first of all admit, with M. Berthollet, that chemical action is exercised indefinitely in a continuous manner between the molecules of substances, whatever their number and ratio may be, and that in general we can obtain compounds with very variable proportions. But then we must admit at the same time that,—apart from insolubility, cohesion, and elasticity, which tend to produce compounds in fixed proportions,—chemical action

---

* [See Alembic Club Reprints, No. 2, p. 41.]
† [See Alembic Club Reprints, No. 2, p. 35.]

is exerted more powerfully when the elements are in simple ratios or in multiple proportions among themselves, and that compounds are thus produced which separate out more easily. In this way we reconcile the two opinions, and maintain the great chemical law, that whenever two substances are in presence of each other they act in their sphere of activity according to their masses, and give rise in general to compounds with very variable proportions, unless these proportions are determined by special circumstances.

Conclusion.

I have shown in this Memoir that the compounds of gaseous substances with each other are always formed in very simple ratios, so that representing one of the terms by unity, the other is 1, or 2, or at most 3. These ratios by volume are not observed with solid or liquid substances, nor when we consider weights, and they form a new proof that it is only in the gaseous state that substances are in the same circumstances and obey regular laws. It is remarkable to see that ammonia gas neutralises exactly its own volume of gaseous acids; and it is probable that if all acids and alkalies were in the elastic state, they would all combine in equal volumes to produce neutral salts. The capacity of saturation of acids and alkalies measured by volume would then be the same, and this might perhaps be the true manner of determining it. The apparent contraction of volume suffered by gases on combination is also very simply related to the volume of one of them, and this property likewise is peculiar to gaseous substances.

3. EXTRACT FROM A NEW SYSTEM OF CHEMICAL PHILOSOPHY. PART II. (1810). BY JOHN DALTON.

(Appendix, pp. 555-559.)

SOME observations on nitric acid, and the other compounds of azote and oxygen, have been made by Gay Lussac, in the 2d vol. of the Mémoires d'Arcueil. He contends that one *measure* of oxygenous gas unites to two *measures* of nitrous gas to form nitric acid, and to three measures to form nitrous acid. Now I have shewn that 1 measure of oxygen may be combined with 1.3 of nitrous gas, or with 3.5, or with any intermediate quantity whatever, according to circumstances, which he seems to allow ; what, then, is the nature of the combinations below 2, and above 3, of nitrous gas ? No answer is given to this ; but the opinion is founded upon an hypothesis that all elastic fluids combine in equal measures, or in measures that have some simple relation one to another, as 1 to 2, 1 to 3, 2 to 3, &c. In fact, his notion of measures is analogous to mine of atoms ; and if it could be proved that all elastic fluids have the same number of atoms in the same volume, or numbers that are as 1, 2, 3, &c. the two hypotheses would be the same, except that mine is universal, and his applies only to elastic fluids. Gay Lussac could not but see (page 188, Part 1. of this work) that a similar hypothesis had been entertained by me, and abandoned as untenable ; however, as he has revived the notion, I shall make a few observations upon it, though I do not doubt but he will soon see its inadequacy.

Nitrous gas is, according to Gay Lussac, constituted of equal measures of azote and oxygen, which, when combined, occupy the same volume as when free. He quotes Davy, who found 44.05 azote, and 55.95 oxygen by weight, in nitrous gas. He converts these into volumes, and finds them after the rate of 100 azote to 108.9 oxygen. There is, however, a mistake in this; if properly reduced, it gives 100 azote to 112 oxygen, taking the specific gravities according to Biot and Arago. But that Davy has overrated the oxygen 12 per cent. he shews by burning potassium in nitrous gas, when 100 measures afforded just 50 of azote. The degree of purity of the nitrous gas, and the particulars of the experiment, are not mentioned. This one result is to stand against the mean of three experiments of Davy, and may or may not be more correct, as hereafter shall appear. Dr. Henry's analysis of ammonia embraces that of nitrous gas also; he finds 100 measures of ammonia require 120 of nitrous gas for their saturation. Now this will apply to Gay Lussac's theory in a very direct manner; for, according to him, ammonia is formed of 1 measure of azote and 3 of hydrogen, condensed into a volume of 2; it follows, then, that 100 ammonia require 75 oxygen to saturate the hydrogen; hence, 120 nitrous gas should contain 75 oxygen, or 100 should contain 62.5, instead of 50. Here either the theory of Gay Lussac, or the experience of Dr. Henry, must give results wide of the truth. In regard to ammonia too, it may farther be added, that neither is the rate of azote to hydrogen 1 to 3, nor is the volume of ammonia doubled by decomposition, according to the experiments of Berthollet, Davy, and Henry, made with the most scrupulous attention to accuracy, to which may be added my own.—There is another point of view in which this theory of Gay Lussac is unfortunate, in regard to ammonia and nitrous gas; 1 measure of azote with 3 of hydrogen, forms 2 of ammonia; and 1 measure of azote with 1 of oxygen, forms 2 of nitrous gas: now,

according to a well established principle in chemistry,
1 measure of oxygen ought to combine with 3 of hydrogen,
or with one half as much, or twice as much ; but no one
of these combinations takes place. If Gay Lussac adopt
my conclusions, namely, that 100 measures of azote
require about 250 hydrogen to form ammonia, and that
100 azote require about 120 oxygen to form nitrous gas,
he will perceive that the hydrogen of the former would
unite to the oxygen of the latter, and form water, leaving
no excess of either, further than the unavoidable errors of
experiments might produce ; and thus the great chemical
law would be preserved. The truth is, I believe, that
gases do not unite in equal or exact measures in any one
instance ; when they appear to do so, it is owing to the
inaccuracy of our experiments. In no case, perhaps, is
there a nearer approach to mathematical exactness, than
in that of 1 measure of oxygen to 2 of hydrogen ; but
here, the most exact experiments I have ever made, gave
1.97 hydrogen to 1 oxygen.

4. ESSAY ON A MANNER OF DETER-MINING THE RELATIVE·MASSES OF THE ELEMENTARY MOLECULES OF BODIES, AND THE PROPORTIONS IN WHICH THEY ENTER INTO THESE COMPOUNDS. By A. AVOGADRO.*

I.

M. GAY-LUSSAC has shown in an interesting Memoir (Mémoires de la Société d'Arcueil, Tome II.) that gases always unite in a very simple proportion by volume, and that when the result of the union is a gas, its volume also is very simply related to those of its components. But the quantitative proportions of substances in compounds seem only to depend on the relative number of molecules † which combine, and on the number of composite molecules which result. It must then be admitted that very simple relations also exist between the volumes of gaseous substances and the numbers of simple or compound molecules which form

---

* Journal de Physique, LXXIII. (1811), pp. 58-76.

† [Avogadro has been accused of inconsistency in his use of the term "molecule," but a careful perusal of his paper will show that he uses it with its qualifying adjectives quite consistently, as follows :—

Molécule (translated " molecule ") without qualification means in modern chemical phraseology either *atom* or *molecule*.

Molécule intégrante (translated " integral molecule ") means *molecule* in general, but is usually applied only to compounds.

Molécule constituante (translated " constituent molecule ") is employed to denote the *molecule* of an elementary substance.

Molécule élémentaire (translated " elementary molecule ") stands for the *atom* of an elementary substance.]

them. The first hypothesis to present itself in this connection, and apparently even the only admissible one, is
the supposition that the number of integral molecules in
any gases is always the same for equal volumes, or
always proportional to the volumes. Indeed, if we were
to suppose that the number of molecules contained in a
given volume were different for different gases, it would
scarcely be possible to conceive that the law regulating
the distance of molecules could give in all cases relations
so simple as those which the facts just detailed compel
us to acknowledge between the volume and the number
of molecules. On the other hand, it is very well conceivable that the molecules of gases being at such a distance
that their mutual attraction cannot be exercised, their
varying attraction for caloric may be limited to condens-
ing a greater or smaller quantity around them, without
the atmosphere formed by this fluid having any greater
extent in the one case than in the other, and, consequently, without the distance between the molecules varying; or, in other words, without the number of molecules
contained in a given volume being different. Dalton, it
is true, has proposed a hypothesis directly opposed to
this, namely, that the quantity of caloric is always the
same for the molecules of all bodies whatsoever in the
gaseous state, and that the greater or less attraction for
caloric only results in producing a greater or less condensation of this quantity around the molecules, and
thus varying the distance between the molecules themselves. But in our present ignorance of the manner in
which this attraction of the molecules for caloric is·
exerted, there is nothing to decide us *à priori* in favour·
of the one of these hypotheses rather than the other;
and we should rather be inclined to adopt a neutral
hypothesis, which would make the distance between the
molecules and the quantities of caloric vary according

to unknown laws, were it not that the hypothesis we have just proposed is based on that simplicity of relation between the volumes of gases on combination, which would appear to be otherwise inexplicable.

Setting out from this hypothesis, it is apparent that we have the means of determining very easily the relative masses of the molecules of substances obtainable in the gaseous state, and the relative number of these molecules in compounds; for the ratios of the masses of the molecules are then the same as those of the densities of the different gases at equal temperature and pressure, and the relative number of molecules in a compound is given at once by the ratio of the volumes of the gases that form it. For example, since the numbers 1.10359 and 0.07321 express the densities of the two gases oxygen and hydrogen compared to that of atmospheric air as unity, and the ratio of the two numbers consequently represents the ratio between the masses of equal volumes of these two gases, it will also represent on our hypothesis the ratio of the masses of their molecules. Thus the mass of the molecule of oxygen will be about 15 times that of the molecule of hydrogen, or, more exactly, as 15.074 to 1. In the same way the mass of the molecule of nitrogen will be to that of hydrogen as 0.96913 to 0.07321, that is, as 13, or more exactly 13.238, to 1. On the other hand, since we know that the ratio of the volumes of hydrogen and oxygen in the formation of water is 2 to 1, it follows that water results from the union of each molecule of oxygen with two molecules of hydrogen. Similarly, according to the proportions by volume established by M. Gay-Lussac for the elements of ammonia, nitrous oxide, nitrous gas, and nitric acid, ammonia will result from the union of one molecule of nitrogen with three of hydrogen, nitrous oxide from one molecule of oxygen with two of nitrogen, nitrous gas from one molecule of nitrogen with one of oxygen, and nitric acid from one of nitrogen with two of oxygen.

II.

There is a consideration which appears at first sight to be opposed to the admission of our hypothesis with respect to compound substances. It seems that a molecule composed of two or more elementary molecules should have its mass equal to the sum of the masses of these molecules; and that in particular, if in a compound one molecule of one substance unites with two or more molecules of another substance, the number of compound molecules should remain the same as the number of molecules of the first substance. Accordingly, on our hypothesis when a gas combines with two or more times its volume of another gas, the resulting compound, if gaseous, must have a volume equal to that of the first of these gases. Now, in general, this is not actually the case. For instance, the volume of water in the gaseous state is, as M. Gay-Lussac has shown, twice as great as the volume of oxygen which enters into it, or, what comes to the same thing, equal to that of the hydrogen instead of being equal to that of the oxygen. But a means of explaining facts of this type in conformity with our hypothesis presents itself naturally enough: we suppose, namely, that the constituent molecules of any simple gas whatever (*i.e.*, the molecules which are at such a distance from each other that they cannot exercise their mutual action) are not formed of a solitary elementary molecule, but are made up of a certain number of these molecules united by attraction to form a single one; and further, that when molecules of another substance unite with the former to form a compound molecule, the integral molecule which should result splits up into two or more parts (or integral molecules) composed of half, quarter, &c., the

number of elementary molecules going to form the constituent molecule of the first substance, combined with
half, quarter, &c., the number of constituent molecules
of the second substance that ought to enter into combination with one constituent molecule of the first substance (or, what comes to the same thing, combined with
a number equal to this last of half-molecules, quarter-
molecules, &c., of the second substance); so that the
number of integral molecules of the compound becomes
double, quadruple, &c., what it would have been if there
had been no splitting-up, and exactly what is necessary to
satisfy the volume of the resulting gas.*

On reviewing the various compound gases most gene-
rally known, I only find examples of duplication of the
volume relatively to the volume of that one of the constituents which combines with one or more volumes of the
other. We have already seen this for water. In the
same way, we know that the volume of ammonia gas is
twice that of the nitrogen which enters into it. M. Gay-
Lussac has also shown that the volume of nitrous oxide
is equal to that of the nitrogen which forms part of it,
and consequently is twice that of the oxygen. Finally,
nitrous gas, which contains equal volumes of nitrogen and
oxygen, has a volume equal to the sum of the two constituent gases, that is to say, double that of each of them.
Thus in all these cases there must be a division of the
molecule into two; but it is possible that in other cases
the division might be into four, eight, &c. The possibility of this division of compound molecules might have
been conjectured *à priori;* for otherwise the integral
molecules of bodies composed of several substances with
a relatively large number of molecules, would come to

---

* Thus, for example, the integral molecule of water will be composed of a half-molecule of oxygen with one molecule, or, what is
the same thing, two half-molecules of hydrogen.

have a mass excessive in comparison with the molecules of simple substances. We might therefore imagine that nature had some means of bringing them back to the order of the latter, and the facts have pointed out to us the existence of such means. Besides, there is another consideration which would seem to make us admit in some cases the division in question ; for how could one otherwise conceive a real combination between two gaseous substances uniting in equal volumes without condensation, such as takes place in the formation of nitrous gas ? Supposing the molecules to remain at such a distance that the mutual attraction of those of each gas could not be exercised, we cannot imagine that a new attraction could take place between the molecules of one gas and those of the other. But on the hypothesis of division of the molecule, it is easy to see that the combination really reduces two different molecules to one, and that there would be contraction by the whole volume of one of the gases if each compound molecule did not split up into two molecules of the same nature. M. Gay-Lussac clearly saw that, according to the facts, the diminution of volume on the combination of gases cannot represent the approximation of their elementary molecules. The division of molecules on combination explains to us how these two things may be made independent of each other.

III.

Dalton, on arbitrary suppositions as to the most likely relative number of molecules in compounds, has endeavoured to fix ratios between the masses of the molecules of simple substances. Our hypothesis, supposing it well-founded, puts us in a position to confirm or rectify his results from precise data, and, above all, to assign the magnitude of compound molecules according to the volumes of the gaseous compounds, which depend partly on the division of molecules entirely unsuspected by this physicist.

Thus Dalton supposes* that water is formed by the union of hydrogen and oxygen, molecule to molecule. From this, and from the ratio by weight of the two components, it would follow that the mass of the molecule of oxygen would be to that of hydrogen as $7\frac{1}{2}$ to 1 nearly, or, according to Dalton's evaluation, as 6 to 1. This ratio on our hypothesis is, as we saw, twice as great, namely, as 15 to 1. As for the molecule of water, its mass ought to be roughly expressed by $15 + 2 = 17$ (taking for unity that of hydrogen), if there were no division of the molecule into two; but on account of this division it is reduced to half, $8\frac{1}{2}$, or more exactly 8.537, as may also be found directly by dividing the density of aqueous vapour 0.625 (Gay-Lussac) by the density of hydrogen 0.0732. This mass only differs from 7, that assigned to it by Dalton, by the difference in the values for the composition of water; so that in this respect Dalton's result is approximately correct from the combination of two compensating errors,—the error in the mass of the molecule of oxygen, and his neglect of the division of the molecule.

Dalton supposes that in nitrous gas the combination of nitrogen and oxygen is molecule to molecule : we have seen on our hypothesis that this is actually the case. Thus Dalton would have found the same molecular mass for nitrogen as we have, always supposing that of hydrogen to be unity, if he had not set out from a different value for that of oxygen, and if he had taken precisely the same value for the quantities of the elements

---

* In what follows I shall make use of the exposition of Dalton's ideas given in Thomson's System of Chemistry. [See Alembic Club Reprints, No. 2, p, 42.]

in nitrous gas by weight. But by supposing the molecule of oxygen to be less than half what we find, he has been obliged to make that of nitrogen also equal to less than half the value we have assigned to it, viz., 5 instead of 13. As regards the molecule of nitrous gas itself, his neglect of the division of the molecule again makes his result approach ours ; he has made it $6 + 5 = 11$, whilst according to us it is about $\dfrac{15 + 13}{2} = 14$, or more exactly

$\dfrac{15.074 + 13.238}{2} = 14.156$, as we also find by dividing 1.03636, the density of nitrous gas according to Gay-Lussac, by 0.07321. Dalton has likewise fixed in the same manner as the facts have given us, the relative number of molecules in nitrous oxide and in nitric acid, and in the first case the same circumstance has rectified his result for the magnitude of the molecule. He makes it $6 + 2 \times 5 = 16$, whilst according to our method it should be $\dfrac{15.074 + 2 \times 13.238}{2} = 20.775$, a number which is also obtained by dividing 1.52092, Gay-Lussac's value for the density of nitrous oxide, by the density of hydrogen.

In the case of ammonia, Dalton's supposition as to the relative number of molecules in its composition is on our hypothesis entirely at fault. He supposes nitrogen and hydrogen to be united in it molecule to molecule, whereas we have seen that one molecule of nitrogen unites with three molecules of hydrogen. According to him the molecule of ammonia would be $5 + 1 = 6$: according to us it should be $\dfrac{13 + 3}{2} = 8$, or more exactly 8.119, as may also be deduced directly from the density of ammonia gas. The division of the molecule, which does not enter into Dalton's calculations, partly corrects in this case also the error which would result from his other suppositions.

All the compounds we have just discussed are produced by the union of one molecule of one of the components with one or more molecules of the other. In nitrous acid we have another compound of two of the substances already spoken of, in which the terms of the ratio between the number of molecules both differ from unity. From Gay-Lussac's experiments (Société d'Arcueil, same volume), it appears that this acid is formed from 1 part by volume of oxygen and 3 of nitrous gas, or, what comes to the same thing, of 3 parts of nitrogen and 5 of oxygen ; whence it would follow, on our hypothesis, that its molecule should be composed of 3 molecules of nitrogen and 5 of oxygen, leaving the possibility of division out of account. But this mode of combination can be referred to the preceding simpler forms by considering it as the result of the union of 1 molecule of oxygen with 3 of nitrous gas, *i.e.* with 3 molecules, each composed of a half-molecule of oxygen and a half-molecule of nitrogen, which thus already includes the division of some of the molecules of oxygen which enter into that of nitrous acid. Supposing there to be no other division, the mass of this last molecule would be 57.542, that of hydrogen being taken as unity, and the density of nitrous acid gas would be 4.21267, the density of air being taken as unity. But it is probable that there is at least another division into two, and consequently a reduction of the density to half : we must wait until this density has been determined by experiment.

IV.

We may now look at a few more compounds, which on our hypothesis can give us at least conjectural information concerning the relative masses of the molecules and their number in these compounds, and compare our results with the suppositions of Dalton.

M. Gay-Lussac has shown that if we assume that dry sulphuric acid is composed of 100 parts of sulphur and 138 of oxygen by weight, as the most recent work of chemists has established, and that the density of sulphurous acid gas is 2.265 referred to air as unity (Kirwan's determination), and if we admit, as the result of Gay-Lussac's experiments, that sulphuric acid is composed of two parts by volume of sulphurous acid gas and one of oxygen, then the volume of sulphurous acid is nearly equal to that of the oxygen which entered into it; and this equality would be exact if the bases on which the calculation rests were the same. If we suppose Kirwan's determination to be exact, throwing the whole error on the analysis of sulphuric acid, we find that in sulphurous acid 100 parts of sulphur take 95.02 of oxygen, and consequently in sulphuric acid

$$95.02 + \frac{95.02}{2} = 142.53,$$ instead of 138. If, on the con-

trary, we suppose the analysis of sulphuric acid to be exact, it follows that sulphurous acid contains 92 of oxygen for 100 of sulphur, and that its specific gravity should be 2.30314, instead of 2.265.

One consideration would appear to weigh in favour of the first assumption, until the density of sulphurous acid gas has been confirmed or rectified by fresh experiments,—namely, that there must have been in the determination of the composition of sulphuric acid, a source of error tending to increase the quantity of the radical, or, what is the same thing, diminish the quantity of oxygen. The determination was made from the quantity of dry sulphuric acid produced. Now it seems almost certain that ordinary sulphur contains hydrogen; the weight of this hydrogen, which must have been converted into water in the operation, has therefore been added to the true weight of the radical. I shall therefore

assume sulphurous acid to be composed of 95.02 * of
oxygen to 100 of sulphur, or rather of sulphuric radical,
instead of 92.†

In order now to determine the mass of the molecule
of the sulphuric radical, it would be necessary to know
what proportion by volume this radical in the gaseous
state would bear to the oxygen in the formation of sulphurous acid. The analogy with other combinations
already discussed, where there is in general a doubling
of the volume or halving of the molecule, leads us to
suppose that it is the same in this case also, *i.e.* that the
volume of the sulphur as gas is half that of the sulphurous
acid, and consequently also half that of the oxygen with
which it combines. On this supposition the density of

sulphur gas will be to that of oxygen as 100 to $\dfrac{95.02}{2}$, or

47.51 ; which gives 2.323 for the density of gaseous
sulphur, taking that of air as unity. The masses of the
molecules being according to our hypothesis in the same
ratio as the densities of the gases to which they belong,
the mass of the molecule of the sulphuric radical will be
to that of hydrogen as 2.323 to 0.07321, or as 31.73 to 1.
One of these molecules combined, as we have said, with
two of oxygen, will form sulphurous acid (division of the
molecule being left out of account), and combined with
yet another molecule of oxygen, will form sulphuric acid.
Accordingly, sulphurous acid should be analogous to

---

* [Erroneously 92.02 in the original.]

† This was written before I had seen the Memoir of Davy on
oxymuriatic acid, which also contains new experiments on sulphur
and phosphorus. In it he determines the density of sulphurous
acid gas, and finds it to be only 2.0967, which gives new force to
the above considerations. If we adopt this density, we find that in
sulphurous acid 100 parts by weight of sulphur take 111 of oxygen,
and in sulphuric acid 167 instead of 138 ; but perhaps this density
of sulphurous acid, according to Davy, is somewhat too low.

nitric acid, with regard to the relative number of molecules of its constituents, sulphuric acid having no analogue amongst the nitrogen compounds. The molecule of sulphurous acid, having regard to division, will be equal to $\dfrac{31.73 + 2 \times 15.074}{2}$, or 30.94, as would also be obtained directly by dividing the density 2.265 of sulphurous acid gas by that of hydrogen gas. As for the molecule of sulphuric acid, it cannot be determined, for we do not know whether there is further division of the molecule on its formation, or not.*

---

* Davy in the Memoir alluded to, has made the same suppositions as to the relative number of molecules of oxygen and radical in sulphurous and sulphuric acids. From his determination of the density of sulphurous acid gas, the density of the sulphuric radical would be 1.9862, and its molecule 27.13, that of hydrogen being taken as unity. Davy, by a similar calculation, fixes it at about half, viz., 13.7, because he supposes the molecule of oxygen to be equal to about half our molecule, using Dalton's hypothesis with respect to water.

He finds nearly the same mass, viz., 13.4, by taking as his starting-point the density of sulphuretted hydrogen, which his experiments make equal to 1.0645, a result only slightly different from Kirwan's, and by assuming that this gas (which contains, as we know, its own volume of hydrogen combined with sulphur) is composed of one molecule of sulphur and one of hydrogen. As we suppose the molecule of sulphur to be nearly twice as great, we must assume that this gas is the product of the union of one molecule of the radical with two at least of hydrogen, and that its volume is twice that of the gaseous radical, as in so many other cases. I say *at least* with two molecules of hydrogen, for if there were hydrogen already in ordinary sulphur, as known experiments on this substance indicate, its quantity also must be added. If, for instance, ordinary sulphur were composed of one molecule of sulphuric radical and one of hydrogen, sulphuretted hydrogen would be composed of three molecules of hydrogen and one of radical. This could be decided by the comparison of the specific gravities of sulphuretted hydrogen and sulphurous acid gas, if both were known exactly. For example, supposing Davy's determination for sul-

Dalton had supposed that sulphuric acid was composed of two molecules of oxygen to one of radical, and
sulphurous acid of one molecule of oxygen to one of
sulphur. These two assumptions are incompatible, for
according to Gay-Lussac's results the quantities of
oxygen in these two acids for a given quantity of radical,
are represented by 1 and 1½. Besides, in his determination of the molecule he set out from a wrong value for
the composition of sulphuric acid, and it is only by
chance that the mass 15 which he assigns to it, bears to
his value for the mass of the oxygen molecule a ratio which
approaches that presented by these two substances on
our hypothesis.

Phosphorus has so much analogy with sulphur that we
might apparently assume that phosphoric acid also is
composed of three molecules of oxygen to one of
radical, and phosphorous acid of only two of oxygen to
one of radical. On this assumption we may calculate
approximately the mass of the molecule of the phosphoric
radical. Rose found by a method analogous to that
which had been employed for sulphuric acid, that phosphoric acid contains about 115 parts by weight of
oxygen to 100 of phosphorus. There ought to be a
little more oxygen in it if we suppose that phosphorus,
like sulphur, contains hydrogen. As an approximation

---

phuretted hydrogen to be exact, the molecule of the sulphuric
radical, on the supposition of only two molecules of hydrogen, would
be 27.08, that of hydrogen being taken as unity; but on the supposition of three molecules of hydrogen, 27.08 would still be the
sum of one molecule of radical and one of hydrogen, so that the
former would be reduced to 26.08. If the exact density of sulphurous acid gas confirmed one or other of these results, it would
confirm by that means one or other of these hypotheses: but we
are not sufficiently agreed about these densities to be able to draw
any conclusion in this respect from the determinations hitherto
existing.

we can make this increase in the same proportion as we have seen holds good for sulphuric acid in accordance with the specific gravity of sulphurous acid, and thus bring the quantity of oxygen up to 120. We then find from our hypotheses that the mass of the molecule of the phosphoric radical is about 38, that of hydrogen being taken as unity. Dalton also has adopted for phosphorous and phosphoric acids, hypotheses analogous to those he had made for sulphurous and sulphuric acids ; but since he used different values for the elements of these acids by weight, he arrived at a determination of the molecule of phosphorus, which does not bear the same ratio to his determination of the molecule of sulphur as ought to exist, according to us, between these molecules : he has fixed that of phosphorus as 8, hydrogen being unity.*

Let us now see what conjecture we may form as to the mass of the molecule of a substance which plays in nature a far greater part than sulphur or phosphorus, namely, that of carbon. As it is certain that the volume of carbonic acid is equal to that of the oxygen which enters into it, then, if we admit that the volume of carbon, supposed gaseous, which forms the other element, is doubled by the division of its molecules into two, as in several combinations of that sort, it will be necessary to suppose that this volume is the half of that of the oxygen with which it combines, and that con-

---

* Davy has adopted the same suppositions as ourselves for the number of molecules of oxygen and radical in phosphorous and phosphoric acids ; and by still taking the molecule of oxygen nearly half ours, he finds 16.5 for the molecule of phosphorus, which would give about 33 on our evaluation of the molecule of oxygen, instead of 38. The difference arises from Davy having taken from his own experiments 34 parts to 25, *i.e.* 136 to 100 of phosphorus, as the quantity of oxygen in phosphoric acid, instead of 120 as we have assumed. Further experiments will clear up this point.

sequently carbonic acid results from the union of one molecule of carbon and two of oxygen, and is therefore analogous to sulphurous and phosphorous acids, according to the preceding suppositions. In this case we find from the proportion by weight between the oxygen and the carbon, that the density of carbon as gas would be 0.832 with respect to that of air as unity, and the mass of its molecule 11.36 with respect to hydrogen. There is, however, one difficulty in this supposition, for we give to the molecule of carbon a mass less than that of nitrogen and oxygen, whereas one would be inclined to attribute the solidity of its aggregation at the highest temperatures to a higher molecular mass, as is observed in the case of the sulphuric and phosphoric radicals. We might avoid this difficulty by assuming a division of the molecule into four, or even into eight, on the formation of carbonic acid; for in that way we should have the molecule of carbon twice or four times as great as that we have just fixed. But such a composition would not be analogous to that of the other acids; and, besides, according to other known examples, the assumption or not of the gaseous state does not appear to depend solely on the magnitude of the molecule, but also on some other unknown property of substances. Thus we see sulphurous acid in the form of a gas at the ordinary temperature and pressure of the atmosphere notwithstanding its large molecule, which is almost equal to that of the solid sulphuric radical. Oxygenated muriatic acid gas has a density, and consequently a molecular mass, still more considerable. Mercury, which as we shall see further on, should have an extremely large molecule, is nevertheless gaseous at a temperature infinitely lower than would be necessary to vaporise iron, the molecule of which is smaller. Thus there is nothing to prevent us from regarding carbonic acid to be composed in the manner indicated above,—and therefore analogous to nitric, sulphuric, and phosphoric acids,— and the molecule of carbon to have a mass expressed by 11.36.

Dalton has made the same supposition as we have done regarding the composition of carbonic acid, and has consequently been led to attribute to carbon a mole-cule equal to 4.4, which is almost in the same ratio to his value for that of oxygen as 11.36 is to 15, the mass of the molecule of oxygen according to us.

Assuming the values indicated for the mass of the molecule of carbon and the density of its gas, carbonic oxide will be formed, according to the experiments of M. Gay-Lussac, of equal parts by volume of carbon gas and oxygen gas ; and its volume will be equal to the sum of the volumes of its constituents : it will accordingly be formed of carbon and oxygen united molecule to mole-cule, with subsequent halving—all in perfect analogy to nitrous gas.

The mass of the molecule of carbonic acid will be—

$$\frac{11.36 + 2 \times 15.074}{2} = 20.75 = \frac{1.5196}{0.07321},$$

and that of carbonic oxide will be—

$$\frac{11.36 + 15.074}{2} = 13.22 = \frac{0.96782}{0.07321}.$$

V.

Amongst the simple non-metallic substances there is still one of which we have to speak. This substance, being naturally gaseous, can leave no doubt, on our principles, as to the mass of its molecule ; but the latest experiments of Davy upon it, and even the earlier experi-ments of MM. Gay-Lussac and Thenard, force us to depart from ideas hitherto received, although the last-

named chemists have attempted to explain them in accordance with these ideas. This is the substance hitherto
known by the name of *oxygenated muriatic acid*, or
oxymuriatic acid. In the present state of our knowledge
we must now regard this substance as still undecom-
posed, and muriatic acid as a compound of it with
hydrogen. It is in accordance with this theory, therefore, that we shall apply our principles regarding combinations to these two substances.

The density of oxymuriatic acid according to MM.
Gay-Lussac and Thenard is 2.470 referred to atmospheric
air as unity ; this gives for its molecule referred to that of
hydrogen as unity, 33.74, adopting the density of hydrogen
determined by MM. Biot and Arago. According to
Davy 100 English cubic inches of oxymuriatic gas
weigh 74.5 grains, and an equal volume of hydrogen gas
2.27. This would give for the molecule of the former
$\frac{74.5}{2.27} = 32.82$. These two estimates differ very little from
the mass that Davy himself, from other considerations,
assigns to this substance, viz., 32.9. It follows from the
experiments both of Gay-Lussac and Thenard, and of
Davy, that muriatic acid gas is formed by the combination of equal volumes of oxymuriatic gas and hydrogen,
and that its volume is equal to their sum. This means,
according to our hypothesis, that muriatic acid is formed
of these two substances united molecule to molecule,
with halving of the molecule, of which we have already
had so many examples. Accordingly the density of
muriatic acid gas, calculating from that given above for
oxymuriatic gas, should be 1.272 ; it is 1.278 according
to the experiments of MM. Biot and Gay-Lussac. If we
suppose this last determination to be exact, the density
of oxymuriatic gas should be 2.483, and the mass of its
molecule 33.91. Should we prefer to adopt this value,

the mass of the molecule of muriatic acid will be
$\frac{34.91}{2} = 17.45 = \frac{1.278}{0.07321}$. The determination of the
specific gravity of the muriatic acid gas by Davy, according to which 100 cubic inches of that gas weigh 39 grains, would give numbers only slightly different, viz., 33.36 for the mass of the molecule of oxymuriatic acid, and 17.18 for that of muriatic acid.

VI.

Let us now apply our hypothesis to some metallic substances. M. Gay-Lussac assumes that mercurous oxide, in the formation of which 100 parts by weight of mercury absorb 4.16 of oxygen, according to Fourcroy and Thenard, is analogous to nitrous oxide, *i.e.*, that the mercury, supposed gaseous, is combined in it with half its volume of oxygen gas, which on our hypothesis is to say that one molecule of oxygen combines with two molecules of mercury. Supposing this to be the case, the density of mercury gas ought to be to that of oxygen as 100 to 8.32, which would give 13.25 as its density taking that of air as unity, and for the mass of the molecule of mercury 181 taking as unity that of hydrogen. On this supposition mercuric oxide, which contains twice as much oxygen, should be formed of mercury and oxygen united molecule to molecule; but some reasons lead me to think that it is mercurous oxide which represents this last case, and that in mercuric oxide one molecule of mercury combines with two of oxygen. Then the density of mercury gas, and the mass of its molecule, would be double what they are on the preceding hypothesis, viz., $26\frac{1}{2}$ for the first, and 362 for the second. In this assumption I am supported by analogies drawn from other metals, and particularly from iron. It follows from the experiments of different chemists, carefully discussed by Hassenfratz, that the two best known oxides of iron, the black oxide and the red oxide, are composed respectively of 31.8 and 45 parts by weight of oxygen to 100 of iron. We see that the second of these two quantities of oxygen is nearly half as great again as the first, so that we are naturally led to suppose that in the first oxide one molecule of iron combines with two molecules of oxygen, and in the second with three. If that is so, and if we admit the proportion for the black oxide to be the more exact, the proportion for the red oxide would be 47.7 for 100 of iron, which comes very near the proportion found directly by Proust, viz., 48. The mass of a molecule of iron will therefore be to the mass of a molecule of oxygen as 100 to 15.9, which gives about 94 with regard to hydrogen as unity. It would appear from this that there should be another oxide of iron which would contain 15.9 of oxygen to 100 of iron, and this is perhaps the white oxide, although the experiments hitherto performed point to this substance containing a greater proportion of oxygen. Now the two oxides of mercury of which we have spoken, one of which contains twice as much oxygen as the other, should apparently be analogous to this last oxide of iron and to the black oxide, the red oxide having no analogue in the case of mercury.

In the same way the other metals present for the most part two oxides in which the quantities of oxygen are as 1 to 2, so that from the proportions of their elements by weight, we may determine in the same manner the mass of their molecules. I find for example, 206 for the molecule of lead, 198 for that of silver, 123 for copper, etc.*

* I shall here add a few words regarding the molecule of potassium. Davy, assuming that potash is formed from potassium and oxygen united molecule to molecule, has fixed the mass of the

VII.

We shall now make a few applications of our principles to saline compounds, which will furnish us with the opportunity of examining an important point in the theory of these compounds. M. Gay-Lussac has shown that the neutral carbonate, fluoborate, and muriate of ammonia are composed of equal volumes of ammonia gas and of the respective acids. Let us pause to consider the carbonate. On our hypothesis, this salt is composed of one molecule of carbonic acid with one molecule of ammonia—*i.e.* (according to the values previously given and independently of any division), of one molecule of

---

molecule of potassium at 40.5, in accordance with the quantity of oxygen by weight in the substance, and has taken the molecule of oxygen to be 7.5. Assuming, as we have done, this last molecule to be nearly twice as great, the molecule of potassium will also be doubled, viz., about 81, if we adopt the other assumptions of Davy. But it may be that in potash one molecule of potassium takes two of oxygen, in which case we should again have to double the last value and make it 162. It might also be (for the analogy drawn from other metals is not in this case a very safe guide) that two molecules of potassium combine with one of oxygen, which would bring back the molecule of potassium to 40.5.

It is on the assumption of this last value for the molecule of potassium that Davy finds 32.9 for that of oxymuriatic acid, calculating from the composition of muriate of potash, and assuming that this salt is formed of one molecule of potassium with one of acid. If we suppose the molecule of potassium to have a different mass, we must admit another relative number of molecules in the muriate, since both from our hypothesis and from the density of the gas, 32.9 is very nearly the molecule of oxymuriatic acid. On the supposition that the molecule of potassium is 81, and that in sulphide of potassium the combination is molecule to molecule, the molecule of sulphur will be about 27 instead of 13½ as found by Davy on the latter assumption, which will bring about between this result and that derived from sulphurous acid according to our calculations the agreement which exists between Davy's values

carbon, two of oxygen, one of nitrogen, and three of hydrogen, which would give 57.75 for the mass of its molecule; but admitting the division into two which had already taken place in the components, this molecule is reduced to 28.87. It would be brought down again to half this number, if there were another division on the union of the acid with the alkali.

M. Gay-Lussac has suspected that the equality of volume between a gaseous alkali and acid, which by their union form a neutral salt, may be general. That is as much as to say, on our hypothesis, that neutral salts are composed of acid and alkali united molecule to molecule; but certain considerations appear to be opposed to the admission of this principle in all its generality. The idea of acidity, alkalinity, and neutrality, which still seems to me the most conformable to the phenomena, is that which I have given in my Memoir on this subject (Journal de Physique, tome lxix.). According to it, all substances form amongst themselves a series, in which they play the part of acid or alkali with respect to one another; and this series is the same as that on which depends the positive or negative electricity they develop on mutual contact. I express by the term *oxygenicity* the property in virtue of which substances are ranked in this scale, placing first those which play the part of an acid with respect to the others. In this scale there is a point about which are placed the substances we term *neutral*, above it are those which are absolutely acid, below it are those which are alkaline, when their state of aggregation permits them to exhibit these qualities. Lastly, composite substances occupy in this scale a place intermediate between those of which they are composed, having regard to the degree of oxygenicity and to the proportion by weight of these constituent substances; so that a neutral substance results from the combination of two substances, one acid, the other alkaline, in a certain proportion (see the Memoir referred to).* The recognition of the simple ratios observed on combination, and in particular in cases where neutral substances are the result, leads us now to a more exact manner of conceiving the state of neutrality. The oxygenicity in two bodies which combine, cannot be supposed to have such a relation to the masses of their molecules, that from the union of certain definite numbers of these molecules there should result a certain definite degree of oxygenicity which would be that of neutrality, and would only depend, as we have already assumed for oxygenicity in general, on the proportion by weight and the degree of oxygenicity of the components. It appears, then, that we must admit that the degree of oxygenicity which corresponds to neutrality is not quite fixed, although approximating more or less to a fixed limit, and that this state depends on the excess of mass of one of the components (from which the acid or alkaline quality might result) being prevented from exercising these qualities by the simple combination with the contrary principle which retains it by its attraction, although the compound otherwise might have a state of aggregation permitting it to act as an acid or an alkali, if it were endowed with these qualities. The excess of mass thus held back is that which is necessary to complete a certain simple relation between the number of combining molecules. Thus amongst the different simple ratios in which molecules can combine, there is one which gives neutrality; that, namely, which gives the

---

* The properties of oxymuriatic acid, as Davy conceives them, being analogous to those of oxygen, are not at all extraordinary from this point of view; they simply show that this substance is very oxygenic. I had already remarked in my Memoir that the properties of the alkalies, supposed to be oxides, are easily explained according to these ideas.

D

compound approximating most closely to the definite point of oxygenicity mentioned above, so that if in the compound formed according to this ratio, one of the component principles let one molecule of the other escape, or took up one in addition, the compound would diverge further from this precise point, about which there oscillate, as it were, the oxygenicities of the various neutral compounds; and it is this point which would give the neutral state in the combination of two substances which could combine in all proportions, or in ratios expressible by any number of molecules whatever. It is evident that this way of regarding the neutrality of compound substances reconciles the theory given in the Memoir quoted with the ideas put forward by M. de Laplace on this point, and expounded by M. Haüy in his Traité de Physique.

According to this theory it is evident that if the oxygenicity of two acids and two alkalies which combine respectively in pairs, is not extremely different, and if at the same time the mass of the molecule of one of the acids is not in a different ratio to its alkali from that of the other acid with regard to its own alkali, the ratio between the numbers of molecules which gives neutrality may be the same in both compounds; but in the contrary case, the ratio may vary in such a way that instead of the equality of volumes, or of combination molecule to molecule which we see between carbonic and a few other acids on the one hand and ammonia on the other, there may be other simple ratios such as 1 to 2, &c., which give the neutral state. Nevertheless, the simplicity which will always exist amongst these ratios, in conjunction with the information we may obtain from other sources as to the mass of the molecules and the degree of oxygenicity of the components, will sometimes put us in a position to determine, or at least conjecture, what are the simple ratios which may occur in a given case ; but it is the task of experiment to confirm or correct these theoretical estimates.

VIII.

It will have been in general remarked on reading this Memoir that there are many points of agreement between our special results and those of Dalton, although we set out from a general principle, and Dalton has only been guided by considerations of detail.   This agreement is an argument in favour of our hypothesis, which is at bottom merely Dalton's system furnished with a new means of precision from the connection we have found between it and the general fact established by M. Gay-Lussac. Dalton's system supposes that compounds are made in general in fixed proportions, and this is what experiment shows with regard to the more stable compounds and those most interesting to the chemist.   It would appear that it is only combinations of this sort that can take place amongst gases, on account of the enormous size of the molecules which would result from ratios expressed by larger numbers, in spite of the division of the molecules, which is in all probability confined within narrow limits. We perceive that the close packing of the molecules in solids and liquids, which only leaves between the integral molecules distances of the same order as those between the elementary molecules, can give rise to more complicated ratios, and even to combinations in all proportions ; but these compounds will be so to speak of a different type from those with which we have been concerned, and this distinction may serve to reconcile M. Berthollet's ideas as to compounds with the theory of fixed proportions.

THE DARIEN PRESS, BRISTO PLACE, EDINBURGH.

ALEMBIC CLUB REPRINTS

OF

IMPORTANT & SCARCE HISTORICAL WORKS

RELATING TO

CHEMISTRY.

Crown Octavo. Cloth. Uniform.

————>◄———

Volumes already Published.

No. 1.—EXPERIMENTS UPON MAGNESIA ALBA, Quick-Lime, and other Alcaline Substances. By JOSEPH BLACK, M.D. 1755. 47 pp. Price 1s. 6d. net.

No. 2.—FOUNDATIONS OF THE ATOMIC THEORY: Comprising Papers and Extracts by JOHN DALTON, WILLIAM HYDE WOLLASTON, M.D., and THOMAS THOMSON, M.D. 1802-1808. 48 pp. Price 1s. 6d. net.

No. 3.—EXPERIMENTS ON AIR. Papers published in the Philosophical Transactions. By the Hon. HENRY CAVENDISH, F.R.S. 1784-1785. 52 pp. Price 1s. 6d. net.

Postage of any of the above to any part of the World,
2d. each extra.

WILLIAM F. CLAY, Publisher,
18 TEVIOT PLACE, EDINBURGH.

www.ingramcontent.com/pod-product-compliance
Lightning Source LLC
Chambersburg PA
CBHW031807090426
42739CB00008B/1197